THE POETRY OF VANADIUM

The Poetry of Vanadium

Walter the Educator™

SKB

Silent King Books a WhichHead Imprint

Copyright © 2023 by Walter the Educator™

All rights reserved. No part of this book may be reproduced in any manner whatsoever without written permission except in the case of brief quotations embodied in critical articles and reviews.

First Printing, 2023

Disclaimer
This book is a literary work; poems are not about specific persons, locations, situations, and/or circumstances unless mentioned in a historical context. This book is for entertainment and informational purposes only. The author and publisher offer this information without warranties expressed or implied. No matter the grounds, neither the author nor the publisher will be accountable for any losses, injuries, or other damages caused by the reader's use of this book. The use of this book acknowledges an understanding and acceptance of this disclaimer.

"Earning a degree in chemistry changed my life!"
- Walter the Educator

dedicated to all the chemistry lovers, like myself, across the world

CONTENTS

Dedication v

Why I Created This Book? 1

One - Vanadium's Might 2

Two - Chemical Art 4

Three - Beauty And Grace 6

Four - Forever Bold 8

Five - Chemistry Of Life 10

Six - Embodiment Of Soul 12

Seven - Element Grand 14

Eight - Alloys It Blends 16

Nine - Universe To Declare 18

Ten - Enchanting And Free 20

Eleven - Conductor Of Energy 22

Twelve - Warrior Of Chemistry 24

Thirteen - Depths Of The Earth 26

Fourteen - Shaping The Land 28

Fifteen - Silent Force 30

Sixteen - Mysterious And Pure 32

Seventeen - Ebb And Flow 34

Eighteen - Adaptation And Fortitude . . . 36

Nineteen - Everlasting Essence 38

Twenty - Laboratories And Fields 40

Twenty-One - A Lasting Trace 42

Twenty-Two - Cosmic Scene 44

Twenty-Three - Steadfast And True . . . 46

Twenty-Four - Flames Of Discovery . . . 48

Twenty-Five - Oh Vanadium 50

Twenty-Six - Power Divine 52

Twenty-Seven - Atomic Sea 54

Twenty-Eight - Crimson Blooms 56

Twenty-Nine - Soaring High 58

Thirty - Luminous Spark 60

Thirty-One - Ores 62

Thirty-Two - Legacy 64

Thirty-Three - Power Of Chemistry 65

Thirty-Four - Crucible Of Creation 67

Thirty-Five - Steel To Sky 69

Thirty-Six - Spark Of Light 71

About The Author 73

WHY I CREATED THIS BOOK?

Creating a poetry book about the chemical element Vanadium is a unique and creative way to explore the characteristics and significance of this element. Poetry has the ability to capture the essence of elements in a lyrical and expressive manner, allowing readers to connect with the scientific and symbolic aspects of Vanadium in a more artistic and emotional way. By delving into the properties, history, and uses of vanadium through poetry, this book can offer a fresh perspective, sparking interest and curiosity in both science enthusiasts and poetry lovers alike.

ONE

VANADIUM'S MIGHT

In the heart of earth's fiery core,
Lies a metal with strength galore.
Vanadium, noble and true,
In crystals, it shines like morning dew.
　　A warrior in the periodic table,
With properties that are truly stable.
It binds with steel, forging it strong,
In the heat of battle, it belongs.
　　Its name derived from a Norse goddess,
Vanadis, her grace and prowess.
Much like her, Vanadium stands tall,
In alloys and catalysts, it conquers all.
　　From the mines to the laboratory,
Its versatility tells a grand story.

In the rust of redox reactions,
It sparks a symphony of chemical attractions.
    So raise a glass to Vanadium's might,
A silent hero, shining bright.
In the realm of elements, it reigns supreme,
Vanadium, a metal of every dream.

TWO

CHEMICAL ART

In the heart of the earth, where fire and metal meet,
Lies Vanadium, a noble and mighty feat.
Named after Norse goddess, strong and divine,
It binds with steel, in alloys it does shine.
    A silent hero, in the depths it dwells,
Conquering all, in catalysts, it excels.
With valiant strength, it weaves its tale,
In colors changing, like a mystical veil.
    In compounds and reactions, it plays its part,
Versatile and bold, with a warrior's heart.
A symbol of power, in the periodic table's scheme,
It sparks the imagination, like a visionary dream.
    Oh Vanadium, metal of every dream,
In chemistry's dance, you reign supreme.

A symbol of might, in a world of extremes,
You stand tall and proud, in the realm of gleams.
 So here's to Vanadium, noble and true,
A metal of wonder, in all that it can do.
In the alchemy of life, you play your part,
Vanadium, the beating heart of every chemical art.

THREE

BEAUTY AND GRACE

In the heart of the earth, Vanadium lies,
A noble metal, radiant and wise.
Its strength and power, a sight to behold,
In alloys and catalysts, its story is told.
    A warrior of elements, fierce and bold,
Conquering compounds, a sight to behold.
In reactions, it dances with grace,
Vanadium, a marvel, in every place.
    A beacon of light, in the chemical art,
Vanadium, the beat of every heart.
A creator of colors, vibrant and true,
In minerals and ores, it shines through.
    A symbol of might, in the periodic line,
Vanadium, a treasure, divine and fine.

In the depths of creation, it plays its part,
Vanadium, the star of every chemical art.
   So let us marvel at this element rare,
In laboratories and fields, beyond compare.
For Vanadium, in its beauty and grace,
Is a wonder of nature, a marvel to embrace.

FOUR

FOREVER BOLD

In the heart of steel, you find me, Vanadium strong,
Forged in the fire, where power and beauty belong.
A catalyst of change, a warrior in disguise,
I bring strength and color, under the endless skies.
 In alloys, I stand tall, enhancing their might,
With resilience and grace, I shine in the light.
I bond with steel, creating blades of might,
A symbol of power, in the darkest night.
 In compounds, I dance, with a vibrant hue,
Painting the world, in colors bold and true.
From red to green, I weave a magical spell,
A spectrum of beauty, where my secrets dwell.
 In the earth's embrace, I quietly lay,
A guardian of strength, in the metal array.

Vanadium, they call me, a force to behold,
In the alchemy of life, a story untold.
    So cherish my presence, in the ores and the ore,
For in my essence, lies a power to explore.
Vanadium, the element, with a tale to be told,
In the chemistry of wonder, forever bold.

FIVE

CHEMISTRY OF LIFE

In the heart of steel, you dwell, Vanadium, bold and bright,
A silent force, a hidden tale, within the metal's might.
Binding with iron, strong and true, you lend your strength and grace,
Forging blades and beams anew, in every time and place.
 In compounds, you reveal your face, with colors rich and rare,
From crimson red to emerald green, a beauty beyond compare.
Oxidation states, a wondrous dance, in reactions swift and sure,
Your versatility, a graceful stance, in the alchemy of nature.

A catalyst of change, you are, in the chemistry of life,
A guardian of strength, near and far, amidst tumult and strife.
Vanadium, oh element rare, your story yet unfolds,
In the symphony of the universe, your mystery still holds.

SIX

EMBODIMENT OF SOUL

In the heart of earth's embrace, Vanadium resides,
A titan among elements, where strength abides.
In compounds and alloys, its power is revealed,
As a force to be reckoned with, its might unconcealed.
    A chameleon of chemistry, it takes on many forms,
From vibrant solutions to solid, steadfast norms.
In the depths of steel, it lends its sturdy hand,
Binding with iron, a union truly grand.
    With a touch of blue in its mineral hue,
Vanadium shines, a spectacle to view.
A gem among elements, it dazzles and gleams,
In the tapestry of nature, it reigns supreme.
    A conductor of energy, it sparks and ignites,
In the dance of electrons, it soars to new heights.

A catalyst of change, it drives reactions forward,
In the alchemy of life, its influence is honored.
 As the sun sets on the horizon of creation,
Vanadium stands tall, a symbol of transformation.
In the symphony of elements, it plays a vital role,
A testament to its resilience, an embodiment of soul.

SEVEN

ELEMENT GRAND

In the heart of the earth, Vanadium lies,
A metal of strength, under darkened skies.
Its valiant spirit, a fiery soul,
In the realm of elements, it holds control.
    With a touch of magic, it paints the scene,
In vibrant hues of blue and green.
A catalyst of change, it sparks the flame,
In chemical dances, it plays the game.
    Partner to iron, in alloys it binds,
Creating swords, with edges that find.
Vanadium's power, steadfast and true,
In the forge of creation, it shines through.
    A warrior of chemistry, bold and bright,
In reactions, it dances with pure delight.

A symbol of beauty, in the periodic chart,
Vanadium's essence, a work of art.
  So let's raise a toast, to this element grand,
In the tapestry of nature, it takes a stand.
Vanadium, oh Vanadium, we sing your praise,
In the alchemy of life, you set the blaze.

EIGHT

ALLOYS IT BLENDS

In the depths of earth, Vanadium lies,
A metal of strength, it does surmise.
Its presence in compounds, a versatile role,
In reactions and catalysts, it plays a vital role.

In minerals and ores, it quietly sleeps,
Symbolizing might, in the secrets it keeps.
A warrior's heart, in the periodic table it stands,
Unyielding and powerful, in the alchemist's hands.

From blades of old to modern steel,
Its mark is felt, its presence real.
With vibrant colors in compounds it shows,
A dazzling display, like a radiant rose.

Bound to iron, in alloys it blends,
A union of strength, as the story extends.

A catalyst of change, in chemical lore,
Vanadium's might, we can't ignore.
   In the periodic chart, it holds its place,
A symbol of power, a figure of grace.
In reactions and compounds, its beauty unfurls,
Vanadium, a metal that conquers and twirls.

NINE

UNIVERSE TO DECLARE

In the heart of steel, you dwell, Vanadium divine,
A catalyst of strength, in every atom you shine.
Your presence in alloys, a secret to behold,
Forged into blades, a story of legends untold.

Beneath the earth, you lie, in colors bright and bold,
A spectrum of beauty, a tale waiting to be told.
From crimson red to emerald green, your hues enchant,
A kaleidoscope of wonder, in every single plant.

In the laboratory, you dance, a catalyst so grand,
Unleashing reactions, with a gentle guiding hand.
Oxidation states, a symphony you compose,
Transforming compounds, where new pathways arose.

Vanadium, oh Vanadium, your mystery runs deep,
A marvel of nature, in secrets you do keep.

In the periodic table, you proudly stand,
A symbol of power, in every grain of sand.
    So here's to you, Vanadium, element so rare,
In your essence, we find, a universe to declare.
A symphony of strength, a palette of grace,
In your boundless spirit, we find our rightful place.

TEN

ENCHANTING AND FREE

In the heart of the earth's embrace,
Lies a metal of strength and grace.
Vanadium, noble and bold,
In compounds, its colors unfold.
    Partner to iron, it forges alloys,
Creating swords that history enjoys.
In the dance of chemical reactions,
It wields power, defying factions.
    A catalyst of change, it leads the way,
Unleashing reactions, come what may.
In the periodic chart, it holds its place,
A mystery, yet to fully embrace.
    With energy, it conducts the flow,
In the alchemy of life, it puts on a show.

Vanadium, enchanting and free,
A vital element, for all to see.

ELEVEN

CONDUCTOR OF ENERGY

In the heart of steel, you shine bright,
Vanadium, element of strength and might.
A catalyst of change, you dance with fire,
In the crucible of creation, you never tire.
    Your atomic dance, a symphony of power,
In compounds and alloys, you boldly tower.
A warrior of chemistry, you never yield,
In colors and forms, your secrets concealed.
    From rusted ore to a gemstone's gleam,
You shape the world with your vibrant dream.
In the depths of earth, you silently reside,
Yet your influence spreads far and wide.
    Oh Vanadium, mysterious and bold,
In the periodic table, your story's told.

A conductor of energy, a force to behold,
In each reaction, your legacy unfolds.
    So here's to you, element of grace,
In laboratories and nature's embrace.
Vanadium, you hold a special place,
In the tapestry of atoms, a marvel of space.

TWELVE

WARRIOR OF CHEMISTRY

In the dance of creation, Vanadium reigns,
A catalyst of change, where mystery remains.
In compounds it weaves, a tapestry so bright,
With hues of green and blue, a mesmerizing sight.
    A titan in transition, it shifts and transforms,
In the alchemy of life, it gracefully performs.
A conductor of energy, it pulses and glows,
In the heart of chemistry, its presence truly shows.
    From the depths of Earth, it emerges with grace,
A symbol of resilience, in every time and space.
Unyielding and strong, in alloys it resides,
A testament to power, where strength abides.
    So let's raise a toast to Vanadium's might,
In the symphony of elements, it shines so bright.

A warrior of chemistry, a force to behold,
Vanadium, oh Vanadium, a story untold.

THIRTEEN

DEPTHS OF THE EARTH

In the heart of creation, Vanadium resides,
A conductor of energy, a force to behold,
With hues of red and silver, it gracefully glides,
Shifting and transforming, a story untold.
    In alloys it dances, with strength and with might,
Resilient and unyielding, in the heat of the flame,
In the alchemy of life, it weaves through the night,
A silent companion, with no need for acclaim.
    From steel to biology, its touch can be found,
A catalyst for change, in the depths of the earth,
A whisper of power, in every sight and sound,
A toast to Vanadium, for all that it's worth.
    So here's to the element, with secrets untold,
A shimmering wonder, in the fabric of time,

May its story be cherished, in silver and gold,
Vanadium's might, in its rhythm and rhyme.

FOURTEEN

SHAPING THE LAND

In the heart of earth's embrace, Vanadium lies,
A prism of hues, a catalyst of change,
With shades of green and blue, it paints the skies,
In alloys and reactions, its role is strange.
    Resilient Vanadium, conductor of might,
Electric dreams flow through its veins,
In circuits and batteries, it shines so bright,
Powering the world with unseen chains.
    Mysterious element, enigma of the deep,
Transformative force, shaping the land,
In steel and industry, its secrets keep,
An enigmatic allure, beyond our command.
    Vanadium, oh Vanadium, element of wonder,
Your presence weaves through time and space,

In chemistry and life, you break asunder,
A marvel of nature, in every place.

FIFTEEN

SILENT FORCE

In the heart of steel, you dwell, Vanadium,
A silent force, a catalyst of change,
Your essence weaves through time's endless loom,
Unyielding, untamed, enigmatic and strange.
    In hues of cerulean, your dreams take flight,
A shimmering dance, a symphony of light,
From sky to sea, your spirit roams,
A whispering echo in nature's ancient tome.
    In alchemy's embrace, you forge new might,
A titan of strength, a guardian of flight,
Mystical and proud, your song resounds,
In every atom, in every bond.
    Through fiery trials and tempests untold,
You stand unbroken, a warrior bold,

Your resilience a testament to time's embrace,
As ages pass, you hold your place.
   Oh Vanadium, in your elemental core,
Lies the secret of creation's evermore,
A symphony of power, a dance of grace,
In the tapestry of existence, you find your place.

SIXTEEN

MYSTERIOUS AND PURE

In the heart of steel and strength you lie,
Vanadium, enigmatic and bold,
A conductor of energy, reaching for the sky,
In alloys and reactions, your story is told.
   Your vibrant hues, a mesmerizing sight,
Shifting from silver to a fiery red,
A chameleon of elements, ever bright,
In the dance of compounds, you confidently tread.
   From batteries to circuits, you shape the world,
Empowering devices with your electric embrace,
A catalyst of change, your power unfurled,
In the symphony of elements, you find your place.
   Resilient and transformative, through time you endure,

A silent force in nature's grand design,
Vanadium, mysterious and pure,
In the tapestry of existence, you brightly shine.

SEVENTEEN

EBB AND FLOW

In the heart of steel, you dwell unseen,
Vanadium, with power serene.
A catalyst of change, you reign,
Transforming all with your vibrant stain.
    In alloys strong, your spirit thrives,
Forging bonds that no force deprives.
Resilient, you stand the test of time,
In every structure, a reason to climb.
    Your hues of green and blue ignite,
A dance of energy, pure and bright.
Conductor of power, you lead the way,
Guiding the currents, night and day.
    In nature's realm, your secrets hide,
A mystery waiting to be untied.

From the depths of earth to the sky above,
Vanadium, you whisper tales of love.
So here's to you, element of might,
Shaping the world with your dazzling light.
Forever bound to the ebb and flow,
Vanadium, in you, we find our glow.

EIGHTEEN

ADAPTATION AND FORTITUDE

In the heart of steel, you gleam, Vanadium,
A whispering catalyst, a silent conductor,
Forged in the core of Earth's fiery crucible,
You dance with the flames, a shimmering enigma.
    Your vibrant hues paint the canvas of creation,
From rusted red to cobalt blue, a spectral symphony,
Unveiling the secrets of your ever-changing soul,
As you weave through the tapestry of transformation.
    In the alchemy of industry, you stand unyielding,
A guardian of strength, a sentinel of resilience,
Empowering the tools of progress with your enduring spirit,
As they carve their mark on the annals of time.
    Mysterious Vanadium, a paradox of stability and

change,
You traverse the realms of science and wonder,
A silent witness to the dance of electrons,
And the unfurling of nature's boundless mysteries.

Oh, elemental titan, your story unfolds in whispers,
An ode to the cosmic forces that birthed you,
A testament to the power of adaptation and fortitude,
Vanadium, eternal wanderer of the periodic table.

NINETEEN

EVERLASTING ESSENCE

In the heart of the earth, Vanadium lies,
A catalyst of change, a conductor of energy,
Its presence sparks transformation, like a phoenix that flies,
In vibrant hues, it dances with vibrant synergy.
    A metal of strength, enduring and bold,
In the crucible of time, its spirit untold,
It weaves through the fabric of existence, unseen,
Yet its influence shapes the world, like a cosmic dream.
    From industrial alloys to the artist's palette,
Vanadium's essence is diverse and malleable,
It paints the canvas of progress with a resolute mettle,

And in the symphony of creation, it plays an indispensable fable.

O Vanadium, in your atomic dance,
You embody resilience, in every circumstance,
A silent force, a mighty presence,
In the tapestry of existence, you're an everlasting essence.

TWENTY

LABORATORIES AND FIELDS

In the heart of steel, your essence blooms,
Vanadium, in vibrant hues.
A conductor of energy, fierce and bold,
Transforming the mundane, into the untold.

In alloys and crystals, your power resides,
Shaping the world, where progress abides.
From deep earth's embrace to the sky's expanse,
Your presence weaves through nature's dance.

Resilient and enigmatic, you stand tall,
In the tapestry of existence, you enthrall.
Mystery veiled in your atomic dance,
Revealing secrets with each fleeting glance.

Oh, Vanadium, in your silent might,
You paint the canvas of day and night.

A whisper in the wind, a spark in the flame,
You etch your mark, without seeking fame.
   So here's to you, element of change,
In laboratories and fields, you rearrange.
A symbol of strength, a force to behold,
Vanadium, in your story untold.

TWENTY-ONE

A LASTING TRACE

In the heart of steel's embrace, you shine,
Vanadium, with hues divine.
A conductor of power's dance,
In you, we find the spark's advance.
    From earth's deep core to sky's expanse,
Your presence weaves a timeless trance.
In nature's grand and wild design,
Your resilience and strength align.
    Mysterious element, enigmatic soul,
In silence, you play a vital role.
A catalyst of change and might,
Shaping the world with silent light.
    In alloys, you endure the test of time,
A silent force in every paradigm.

Adaptable, yet steadfast and true,
Vanadium, we owe much to you.
    In industry's forge and nature's grace,
You leave an imprint, a lasting trace.
So here's to you, element of might,
Vanadium, in your silent flight.

TWENTY-TWO

COSMIC SCENE

In the heart of steel, you gleam and shine,
Vanadium, your spirit so divine.
A silent force, a catalyst of change,
In the dance of atoms, you rearrange.
    Your vibrant hues, a symphony of light,
In alloys and metals, you take flight.
Empowering progress, forging the new,
In the core of innovation, you breakthrough.
    Mysterious element, elusive and bold,
In the depths of Earth, your story's told.
Resilient and strong, you stand the test of time,
A symbol of endurance, in rhythm and rhyme.
    Nature's embrace, you weave your way,
In the wings of birds, in the sky's display.
A touch of magic, in the leaves of green,
In the whispers of wind, your presence keen.

Vanadium, you shape the world unseen,
A quiet hero, in the cosmic scene.
From industry to nature, your legacy unfurls,
In the tapestry of existence, you leave lasting swirls.

TWENTY-THREE

STEADFAST AND TRUE

In the heart of steel, you quietly reside,
Vanadium, a force, strong and dignified.
Your essence infuses the metal's core,
Granting it strength, resilience galore.

In crimson blooms, you paint the earth,
A whisper of life, a symbol of rebirth.
A catalyst of change, in nature's grand design,
You shape the world, with a touch divine.

Amidst the stars, you shimmer and gleam,
A celestial dancer, in the cosmic stream.
Your mysteries unfold, in the nebula's embrace,
An enigma of creation, a shimmering grace.

Vanadium, oh element of might,
In laboratories, you spark invention's light.

From alloys to batteries, you pave the way,
A silent guardian, in the modern age's play.
    So here's to you, oh vanadium rare,
A titan of industry, earth, and air.
In every atom, in every hue,
Your presence endures, steadfast and true.

TWENTY-FOUR

FLAMES OF DISCOVERY

In the heart of steel, you quietly reside,
Vanadium, a silent force, standing by with pride.
Your presence weaves through the metal's core,
Granting strength and resilience forevermore.
　In the cosmic dance, you shimmer and glow,
A celestial dancer with an enigmatic flow.
Mystery cloaks you in an ethereal sheen,
As if you hold the secrets of the universe unseen.
　In nature's embrace, you paint the earth,
In hues of green and gold, celebrating rebirth.
A catalyst for change, you whisper in the wind,
Guiding evolution with the wisdom you've pinned.
　In laboratories, your potential unfurls,
Catalyzing progress, as innovation whirls.

A visionary element, you shape the unknown,
Igniting the flames of discovery, brightly shown.
    Vanadium, enigmatic and bold,
Your essence weaves through stories untold.
A symbol of adaptability, a beacon of might,
You journey through time, a cosmic dream in flight.

TWENTY-FIVE

OH VANADIUM

In the heart of stars, Vanadium gleams,
A cosmic dream, a dancer in the celestial streams.
It weaves through time, a touch of magic rare,
Shaping the world with its silent, steadfast flair.

In nature's embrace, it whispers of life,
A catalyst for growth amid struggle and strife.
Resilient and strong, it stands the test of time,
A symbol of endurance, a spirit sublime.

In industry's forge, it sparks innovation's fire,
A force of progress, lifting humanity higher.
Its presence profound, in alloys and steel,
Transforming the world with its unyielding zeal.

Oh Vanadium, enigmatic and bright,
You paint the tapestry of existence with your light.

A silent force, with mysteries untold,
You leave an indelible mark, a story yet unfold.
    So dance on, celestial dancer of might,
We marvel at your essence, in the day and the night.
A whisper of wonder, a promise untamed,
Vanadium, in our hearts, your legacy is famed.

TWENTY-SIX

POWER DIVINE

In the heart of the earth, Vanadium lies,
A silent force, hidden from prying eyes.
A metal of strength, in nature's grand design,
Shaping the world with a power divine.

From ancient times to the modern age,
Vanadium's presence, an enduring stage.
In ores and rocks, it quietly dwells,
A testament to the stories it tells.

Resilient and bold, it weathers the storm,
Forging new paths, in its steadfast form.
A catalyst of change, in the alchemy of life,
Vanadium's essence, free from strife.

In industry and innovation, it plays its part,
Empowering progress with a fiery heart.

An enigmatic element, mysterious and rare,
Vanadium's allure, beyond compare.
  So here's to Vanadium, so steadfast and true,
A force of nature, in skies of azure blue.
A symbol of adaptability, in a world that's ever-changing,
Vanadium's legacy, eternally engaging.

TWENTY-SEVEN

ATOMIC SEA

In the heart of steel, you quietly reside,
Vanadium, element of strength and pride.
A dancer in the cosmic ballet, you sway,
With grace and poise, in the metal's array.

Your presence weaves through alloys bold,
Infusing them with a resilience untold.
A catalyst for change, you spark the flame,
Innovating, evolving, never the same.

Mysterious whispers in the chemical dance,
You defy confinement, refusing to glance.
Unyielding in nature, you endure the test,
A symbol of adaptability at its best.

From industry's embrace to the earth's deep core,
Your enigmatic allure leaves us wanting more.
A force of progress, you push the bounds,
Unveiling possibilities in leaps and bounds.

Vanadium, oh celestial dancer of might,
In your silent ways, you shine so bright.
A whisper of life in the atomic sea,
Forever entwined with our destiny.

TWENTY-EIGHT

CRIMSON BLOOMS

In the heart of the earth, Vanadium resides,
A catalyst of change, where innovation abides.
With strength and luster, it gleams and shines,
A symbol of resilience, in nature's grand designs.

In crimson blooms, it paints the desert sand,
A fiery hue, spreading across the land.
An enigmatic force, both fierce and bold,
Vanadium, a story waiting to be told.

In steel's embrace, it lends its mighty hand,
Forging the future, where progress takes its stand.
A silent hero, in the industry's grand dance,
Vanadium, the architect of fate's advance.

In alloys and compounds, its magic unfurls,
A conductor of dreams, as the world whirls.
A silent witness to time's unfading trace,
Vanadium, an emblem of enduring grace.

So let us cherish this element of might,
For in its essence, we find courage to fight.
Vanadium, a symbol of hope and change,
In every atom, a story so strange.

TWENTY-NINE

SOARING HIGH

In the heart of the earth, where flames dance bright,
Lies Vanadium, a metal of resolute might.
A catalyst of change, a force so pure,
It sparks the fire of progress to endure.

In alloys and steels, it finds its home,
Granting strength and resilience, where it roams.
A silent guardian, steadfast and bold,
Shaping the future, in ways untold.

In the alchemy of creation, it plays its part,
Mingling with elements, igniting the art.
A mystery unraveling, in its atomic dance,
Vanadium, the enigmatic force of chance.

From rusted ores to gleaming blades,
It weaves its magic in the blacksmith's raids.

A symbol of transformation, it stands tall,
Vanadium, the alchemist of them all.
   So let us honor this element so rare,
For in its essence, we find the power to dare.
To dream, to build, to reach for the sky,
Vanadium, the spark of change, soaring high.

THIRTY

LUMINOUS SPARK

In the heart of steel, you quietly reside,
Vanadium, element of strength and pride.
Catalyst of change, in the furnace you dance,
Transforming the ordinary with a daring stance.
    In the depths of time, your story unfolds,
A silent witness to history, untold.
From swords to skyscrapers, you've left your mark,
In the fabric of progress, a luminous spark.
    Mysterious metal, with a lustrous sheen,
In the alchemy of life, an enigma unseen.
Resilient and unyielding, in the face of the unknown,
You stand as a symbol, of endurance, alone.
    In the laboratory of dreams, you inspire the bold,
Igniting innovation, in a world untold.

Boundless potential, in your atomic core,
Vanadium, element of wonder, forevermore.

    So here's to you, vanadium, in your silent grace,
A testament to the marvels of time and space.
In the tapestry of existence, your legacy weaves,
A reminder of the power, that every atom breathes.

THIRTY-ONE

ORES

In the heart of the earth, Vanadium sleeps,
A silent guardian in the rocks it keeps.
Unyielding and strong, in ores it lies,
A shimmering presence beneath the skies.

Adaptable soul, it weaves through the land,
A catalyst for progress, with a mighty hand.
In steel it finds purpose, giving strength and grace,
An alchemist's dream, in every time and place.

Vanadium, oh Vanadium, enigmatic and bold,
Your secrets untold, yet your story is told.
In colors of beauty, you paint the world bright,
A silent conductor of innovation's flight.

From ancient times to the modern day,
You've stood the test, in every way.
A legacy enduring, beyond measure or time,
Vanadium, you lift humanity higher, in your prime.

So here's to you, element of might,
In the dance of creation, you shine so bright.
A symbol of resilience, in the grand design,
Vanadium, oh Vanadium, your essence divine.

THIRTY-TWO

LEGACY

In the heart of steel, you quietly reside,
Vanadium, a silent guardian with grace,
Shaping the future, where worlds collide,
A touch of alchemy in every embrace.

Your presence, a mystery, lustrous sheen,
Inspiring innovation, a beacon of light,
In the depths of earth, where you've been,
Symbolizing endurance, beyond our sight.

In the forge of creation, you play your part,
Melding strength and grace in every form,
A metal of power, with a beating heart,
Vanadium, in you, we find a calm.

So here's to you, element of might,
In steel and science, you shine so bright,
Vanadium, a legacy, enduring and true,
In every atom, we find strength in you.

THIRTY-THREE

POWER OF CHEMISTRY

In the heart of steel, you quietly reside,
Vanadium, a force, impossible to hide.
In the forge's fiery dance, you play your part,
Infusing strength and grace, a work of art.

A guardian of progress, silent and strong,
In the alloy's embrace, where you belong.
With endurance unmatched, you shape the future,
In every structure, a resilient suture.

Mysterious element, enigmatic and bold,
In the earth's embrace, your story is told.
Your presence in creation's alchemy,
Inspires innovation, a symphony.

Vanadium, your legacy will endure,
In the symphony of steel, forever pure.
A symbol of strength, grace, and might,
Shaping the world, in the morning light.

So, we honor you, in every stride,
Vanadium, our silent guide.
In the fabric of time, you'll always be,
A testament to the power of chemistry.

THIRTY-FOUR

CRUCIBLE OF CREATION

In the heart of steel, you lie, Vanadium,
Forged in the core of Earth's fiery womb,
Your strength weaves through the metal's grain,
A silent force that shapes the future's frame.

In the crucible of creation, you dance,
Mingling with iron, a harmonious trance,
Empowering blades, bridges, and machines,
With resilience that surpasses mortal means.

Your atomic dance, a symphony of might,
In the alchemy of life, a guiding light,
Coursing through cells, a catalyst unseen,
Fueling the fire of innovation's dream.

Vanadium, oh element of enduring grace,
Your presence in history, we can't erase,

A testament to the earth's enduring lore,
As we strive to reach for worlds unknown.
    So, here's to you, Vanadium, steadfast and true,
In every atom, in every breakthrough,
Your legacy echoes through time's endless throng,
A testament to strength, resilience, and song.

THIRTY-FIVE

STEEL TO SKY

In the forge of creation, Vanadium lies,
A catalyst of change, a shimmering prize.
A metal of might, with secrets untold,
Innovating the future, a story unfold.

Its atomic dance, a symphony rare,
Binding the elements with an enigmatic flair.
Resilient and bold, in the crucible's heat,
Vanadium endures, its destiny complete.

From alchemical dreams to modern-day lore,
It whispers of power, of alchemy's core.
In the tapestry of life, its presence is clear,
An element of wonder, devoid of fear.

From steel to sky, it shapes our world,
A muse for inventors, its marvels unfurled.
Inspiring the mind, with a radiant spark,
Vanadium's legacy lights up the dark.

So let us behold this element divine,
A titan of science, in every design.
For in its essence, we find the key,
To unlock the wonders that are yet to be.

THIRTY-SIX

SPARK OF LIGHT

In the heart of steel, you quietly reside,
Vanadium, in strength and grace allied,
Forged in the earth's enduring flame,
Your presence echoes an ancient acclaim.
    A symbol of power, unwavering and bold,
In alchemy's tale, your mystique is told,
A catalyst for change, a spark of light,
Guiding the seekers through the darkest night.
    In the crucible of time, you hold your ground,
Unyielding, unchanging, in silence profound,
Fueling innovation, igniting the forge,
In the quest for progress, you emerge.
    In hues of silver, a shimmering dance,
You grace the world with a timeless expanse,

A testament to resilience, through trials untold,
Vanadium, your story will forever unfold.
   So here's to you, in all your enigmatic flair,
A silent guardian, beyond compare,
In the symphony of elements, you take your place,
Vanadium, an enduring marvel, in time and space.

ABOUT THE AUTHOR

Walter the Educator is one of the pseudonyms for Walter Anderson. Formally educated in Chemistry, Business, and Education, he is an educator, an author, a diverse entrepreneur, and he is the son of a disabled war veteran. "Walter the Educator" shares his time between educating and creating. He holds interests and owns several creative projects that entertain, enlighten, enhance, and educate, hoping to inspire and motivate you.

Follow, find new works, and stay up to date
with Walter the Educator™
at WaltertheEducator.com

www.ingramcontent.com/pod-product-compliance
Lightning Source LLC
LaVergne TN
LVHW010602070526
838199LV00063BA/5051